Readers are encouraged to go to www.MissionPointPress.com to contact the author or to find information on how to buy this book in bulk at a discounted rate.

ISBN: 978-1-943995-46-2

Printed in the United States of America.

Introduction

In 1999, I became unemployed from an automotive manufacturing plant due to a plant closure. That summer I decided to go to the Traverse City Cherry Festival located in Traverse City, Michigan to watch The Navy Blue Angels perform.

The Navy Blue Angels perform during the Traverse City Cherry Festival in Traverse City, Michigan every other year. Since 1999, I have been fortunate to be able to photograph The Navy Blue Angels.

I knew that if I was going to photograph The Navy Blue Angels I would need a better camera then the one I currently owned. I went to what used to be The Blue Photo camera shop in Traverse City, Michigan and purchased a Minolta StSi Maxxium camera and two lens: a AF Zoom 35-80 and a AF Zoom 75-300.

This was my first year photographing The Navy Blue Angels and learning to operate a new camera. This was an expensive lesson—more visible blue sky than anything else. I was lucky to have 10 nice clear photos out of 4 rolls of 36 exposures with a total of 144 photo at a cost of $64.00 dollars to develop and the price of a new camera and lens. That is a 7% outcome of nice photos. Not very good odds.

I have gotten better since 1999. Here are some photos of The Navy Blue Angels from my collection that I would like to share with you. I hope you enjoy these photos of The Navy Blue Angels as much as I enjoyed photographing The Navy Blue Angels.

Grand Traverse Bay

Information from Wikipedia. org

Grand Traverse Bay is a bay of Lake Michigan formed by the Leelanau Peninsula in the Northwestern Lower Peninsula of Michigan. The bay is 32 miles (km) long, 10 mi (16km) wide, and up to 620 feet (190m) deep in spots. It is divided into two arms by the Old Mission Peninsula. The entire bay has a common boundary with the Grand Traverse Bay Bottomland Preserve. It should not be confused with Grand Traverse Bay of Lake Superior, located on the Keweenaw Peninsula.

Grand Traverse Bay includes surrounding Grand Traverse County, Michigan, Antrim County, Michigan and Leelanau County, Michigan.

Traverse City is situated at the south end of the bay where the Boardman River empties into the west arm. Cherry orchards line the bay region, giving rise to Traverse City's claim to be the Cherry Capital of the World. The region is the center of cherry production in Michigan.

The Grand Traverse Bay includes an East and West arm and several important smaller bays: Northport Bay, Suttons Bay, Omena Bay, Bowers Harbor and Old Mission Bay. Northport Bay located at the northwest corner of Grand Traverse Bay is about 10 mi (16 km) long and 4 mi (6.4 km) side. Northport Bay opens to the East, except inside the arms at each end, with the Leelanau Peninsula on the west side.

The Navy Blue Angles is The United States Navy's Flight demonstration squadron, with aviators from The Navy and Marines. The Navy Blue Angels team was formed in 1946, making it the second oldest formal flying aerobatic team (under the same name) in the world, after The French Patrouille de France formed in 1931.

The Navy Blue Angels' six demonstration pilots currently fly the McDonnell Douglas F/A-18 Hornet, typically in more than 70 shows at 34 locations though out The United States each year, where they still employ many of the same practices and techniques used in their aerial displays in their founded 1946 season.

The mission of The United States Navy Flight Demonstration Squadron is "To showcase the pride and professionalism of The United States Navy and Marine Corps by inspiring a culture of excellence and serve to country through flight demonstrations and community outreach."

The parameters of each show must be tailed in accordance with local weather conditions at show time: In clear weather the high show is performed; in overcast conditions a low show is performed, and in limited visibility (weather permitting) the flat show is presented. The high show requires at least an 8,000-foot (2,400m) ceiling and visibility of at least 3 nautical miles (6km) from the show's center point. The minimum ceilings allowed for low and flat shows are 3,500 feet (1km) and 1,500 feet (460m), respectively.

Note: Ceiling is the height above the ground of the base of the lowest layer of clouds.

The Navy Blue Angels Timeline

From Wikipedia.org

1940's

When initially formed, the unit was called The Navy Flight Exhibition Team. The squadron was officially re-designated as The United States Navy Flight Demonstration Squadron in December 1974. The original team was christened The Blue Angels in 1946, when one of the pilots came across the name of New York City's Blue Angel Nightclub in the New Yorker magazine; the team introduced themselves as the "Blue Angels" to the public for the first time on 21 July 1946 in Omaha, Nebraska.

On 25 August 1946 the squadron upgraded their aircraft to the Grumman F8F-1 Bearcat in May 1947. Flight Leader Lt. Cmdr. Bob Clarke replace Butch Voris as the leader of the team and introduced The Famous Diamond Formation, now considered the Blue Angels' trademark.

The official Blue Angels insignia was designed by then team leader Lt. Cmdr. R.E. "Dusty" Rhodes and Virginia Porter (illustrator for Naval Air Advanced Training Command), then approved by Chief of Naval Operations in 1949.

1950's

The "Blue" continued to perform nationwide until the start of the Korean War in 1950, when (due to shortage of pilots, and no planes were available) the team was disbanded and its members were ordered to combat duty.

1970's

The "Blues received their first U.S. Marine Corps Lockheed KC-130F Hercules in 1070. An all-Marine crew manned it."

1980's

One 8 November 1986, The Blue Angels completed their 40th anniversary year during ceremonies unveiling their present aircraft, the McDonnell Douglas F/A 18 Hornet, the first multi-role fighter / attack aircraft. The power and aerodynamics of the Hornet allows then to perform a slow, high angle of attack "tailsitting" maneuver, and to fly a "dirty" (landing gear down) formation loop. The last is not duplicated by the USAF Thunderbirds.

1990's

In 1998, Cmdr. Patrick Driscoll make the first "Blue Jet" landing on a "haze grey and underway" aircraft carrier, USS Harry S. Truman (CVN-75).

Note: Haze grey is a paint color scheme used by USN warships to make the ships harder to see clearly. The grey color reduces the contrast of the ship with the horizon, and reduces the vertical patterns in the ship's appearance.

Underway, as supposed to being anchored, docked alongside, moored or otherwise attach to a fixed place.

"Haze grey and underway" Is shorthand for Naval Surface Warship at sea.

2000's

In 2006, The Blue Angels marked their 60th year of performed.

In 2009, The Blue Angels were inducted into the international Air & Space Hall of Fame at the San Diego Air & Space Museum.

The 2016 season marked the 70th anniversary of an organization that has performed for millions of fans since 1946.

For the last 70 years, a positive mental attitude, determination, and a willingness to learn have been essential attributes of All Blue Angels. Day in and Day out dedication, teamwork, and trust are the keys to this squadron's enduring success.

8

34

39

46

48

50

54

Accidents

From Wikipedia.com

During its history, 27 Blue Angels pilots have been killed in airshow or training accidents. Through the 2006 season there have been 262 pilots in the squadron's history, giving the job a 10% fatality rate.

The last fatality was on 2 June 2016, when Capt. Jeff "Kooch" Kuss, (Opposing Solo, Blue Angel No.6), died just after takeoff while performing the Split-S maneuver in his F/A 18 Hornet during a practice run for The Great Tennessee Air Show in Smyrna, Tennessee. The Navy investigation found that Capt. Kuss performed the maneuver at too low of an altitude while failing to regard the throttle out of afterburner, causing him to fall to fast and recover at too low of an altitude. Capt. Kuss ejected, but his parachute was immediately engulfed in flames, causing him to fall to this death. Kuss' body was recovered multiple yards away from the crash site. The cause of death was blunt force trauma to the head. The investigation also cites weather and pilot fatigue as additional causes to the crash.

In a strange twist, Captain Kuss' fatal crash happened hours after the Blue Angels' fellow pilots in the United States Air Force Thunderbirds suffered a crash of their own following The United States Air Force Academy graduation ceremony earlier that day.

The 2016 Traverse City Cherry Festival in Traverse City, Michigan was the first air show performance since weeks of being grounded following the crash of Captain Kuss, USMC.

Capt. Jeff Kuss, USMC
Opposing Solo

Captain Jeff Kuss is a native of Durango, Colorado, and graduated from Durango High School in 2002. He attended Fort Lewis College, Durango, Colorado, and graduated with a Bachelor of Arts Degree in Economics in 2006. Jeff was commissioned a Second Lieutenant in the U.S. Marine Corps through the Officer Candidate Course in 2006 and reported to The Basic School at Marine Corps Base Quantico, Virginia to complete training.

Captain Kuss joined the Blue Angels in September 2014. He has accumulated more than 1,400 flight hours and 175 carrier-arrested landing. His decorations include The Strike Flight Air Medal, The Navy and Marine Corps Achievement Medal and various personal and unit Awards.

From the Blue Angels Navy Flight Demonstration Squadron brochure

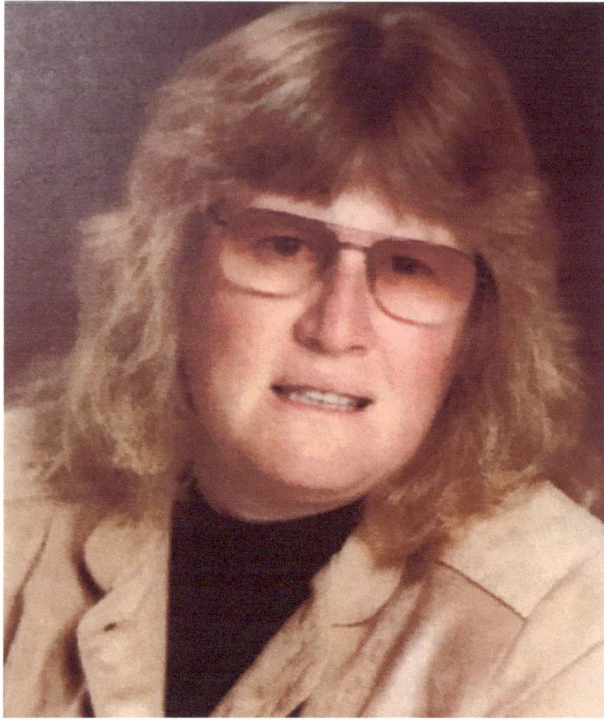

Delores Witkoski, a self-taught photographer with years of experience and a vast photo collection, has earned many degrees including as Associate in Arts and Business Administration from Northwestern Michigan College, a Certificate in International Business and a Bachelor in Business Administration from Ferris State University, and a Master of Science in Administration from Central Michigan University. Delores has also attended Fred Pryor CareerTrack seminars, NMC-EES classes, and "Score" seminars. Delores's teachings and life experiences have given her a broad array of views, while her patience and persistence has allowed her to catch unique moments as they unfold.

Delores dedicates this book to her mother and father for the graduation gift of a camera, her maternal grandparents for the graduation gift of traveling luggage and the opportunity to explore different cultures in Leelanau County, and her paternal grandparents for showing her the teachings of being raised in a Polish culture. She also wants to include long-time friends of 30 years-plus—Tom and Annette, with daughter Kristy and son Michael. Finally, she dedicates this book to her brother and two nephews.

Delores's words of advice: "Do your best and always stay focused on your goals!"

Delores wishes to thank The National Cherry Festival Committee and all who are involved in bringing The Navy Blue Angels to the Traverse City area, without these people this publication would not be possible.

A Special Thank You to the Following:

United Technologies Corporation/Scholar Program/Work Experience

The law office of Bishop & Heintz, P.C.

The law office of Gerald Chefalo

"Score" seminars and mentors

Walgreens' Photo Development Center

SNAP! Printing

Suttons Bay Bingham District Library

Ticconi's ATA of Suttons Bay

Maple City – Glen Lake Community Schools

Northwestern Michigan College

Ferris State University

Central Michigan University

Fred Pryor CarrerTrack seminars

Northwestern Michigan College Extended Educational Services (NMC-EES)

www.ingramcontent.com/pod-product-compliance
Lightning Source LLC
Chambersburg PA
CBHW041100210326

41597CB00004B/140